DIFFICULTIES OF DEVELOPMENT
AS APPLIED TO MAN

- Popular Science Monthly Volume 10 -
November 1876

BY

ALFRED RUSSEL WALLACE

British Library Cataloguing-in-Publication Data
A catalogue record for this book is available from the
British Library

Alfred Russel Wallace

Alfred Russel Wallace was born on 8th January 1823 in the village of Llanbadoc, in Monmouthshire, Wales.

At the age of five, Wallace's family moved to Hertford where he later enrolled at Hertford Grammar School. He was educated there until financial difficulties forced his family to withdraw him in 1836. He then boarded with his older brother John before becoming an apprentice to his eldest brother, William, a surveyor. He worked for William for six years until the business declined due to difficult economic conditions.

After a brief period of unemployment, he was hired as a master at the Collegiate School in Leicester to teach drawing, map-making, and surveying. During this time he met the entomologist Henry Bates who inspired Wallace to begin collecting insects. He and bates continued exchanging letters after Wallace left teaching to pursue his surveying career. They corresponded on prominent works of the time such as Charles Darwin's *The Voyage of the Beagle* (1839) and Robert Chamber's *Vestiges of the Natural History of Creation* (1844).

Wallace was inspired by the travelling naturalists of the day and decided to begin his exploration career collecting specimens in the Amazon rainforest. He explored the Rio Negra for four years, making notes on the peoples and languages he encountered as well as the geography, flora, and fauna. On his return voyage his ship, Helen, caught fire and he and the crew were stranded for ten days before being picked up by the Jordeson, a brig travelling from Cuba to London. All of his specimens aboard Helen had been lost.

After a brief stay in England he embarked on a journey to the Malay Archipelago (now Singapore, Malaysia, and Indonesia). During this eight year period he collected more than 126,000 specimens, several thousand of which represented new species to science. While travelling, Wallace refined his thoughts about evolution and in 1858 he outlined his theory of natural selection in an article he sent to Charles Darwin. This was published in the same year along with Darwin's own theory. Wallace eventually published an account of his travels *The Malay Archipelago* in 1869, and it became one of the most popular books of scientific exploration in the 19th century.

Upon his return to England, in 1862, Wallace became a staunch defender of Darwin's landmark work *On the Origin of Species* (1859). He wrote responses to those critical of the theory of natural selection, including 'Remarks on the Rev. S. Haughton's Paper on the Bee's Cell, And on the Origin of Species' (1863) and 'Creation by Law' (1867). The former of these was particularly pleasing to Darwin. Wallace also published important papers such as 'The Origin of Human Races and the Antiquity of Man Deduced from the Theory of 'Natural Selection" (1864) and books, including the much cited *Darwinism* (1889).

Wallace made a huge contribution to the natural sciences and he will continue to be remembered as one of the key figures in the development of evolutionary theory.

Wallace died on 7th November 1913 at the age of 90. He is buried in a small cemetery at Broadstone, Dorset, England.

DIFFICULTIES OF DEVELOPMENT AS APPLIED TO MAN.[1]

Popular Science Monthly Volume 10 November 1876

AS my own knowledge of and interest in anthropology are confined to the great outlines, rather than to the special details of the science, I propose to give a very brief and general sketch of the modern doctrine as to the Antiquity and Origin of Man, and to suggest certain points of difficulty which have not, I think, yet received sufficient attention.

Many now present remember the time (for it is little more than twenty years ago) when the antiquity of man, as now understood, was universally discredited. Not only theologians, but even geologists, then taught us that man belonged altogether to the existing state of things; that the extinct animals of the Tertiary period had finally disappeared, and that the earth's surface had assumed its present condition, before the human race first came into existence. So prepossessed were even scientific men with this idea—which yet rested on purely negative evidence, and could not be supported by any arguments of scientific value—that numerous facts which had been presented at intervals for half a century, all tending to prove the existence of man at very remote epochs, were silently ignored; and, more than this, the detailed statements of three distinct and careful observers were rejected by a great scientific society as too improbable for publication, only because they proved (if they were true) the coexistence of man with extinct animals![2]

But this state of belief in opposition to facts could not long continue. In 1859 a few of our most eminent geologists examined for themselves into the alleged occurrence of flint implements in the gravels of the north of France, which had been made public fourteen years before, and found them strictly correct. The caverns of Devonshire were about the same time carefully examined by equally eminent observers, and were found fully to bear out the statements of those who had published their results eighteen years before. Flint implements began to be found in all suitable localities in the south of England, when carefully searched for, often in gravels of equal antiquity with those of France. Caverns, giving evidence of human occupation at various remote periods, were explored in Belgium and the south of France—lake-dwellings were examined in Switzerland refuse-heaps in Denmark and thus a whole series of remains have been discovered, carrying back the history of mankind from the earliest historic periods to a long-distant past. The antiquity of the races thus discovered can only be generally determined by the successively earlier and earlier stages through which we can trace them. As we go back, metals soon disappear, and we find only tools and weapons of stone and of bone. The stone weapons get ruder and ruder; pottery, and then the bone implements, cease to occur; and in the earliest stage we find only chipped flints, of rude design, though still of unmistakably human workmanship. In like manner domestic animals disappear as we go backward; and, though the dog seems to have been the earliest, it is doubtful whether the makers of the ruder flint implements of the gravels possessed even this. Still more important as a measure of time are the changes of the earth's surface of the distribution of animals and of climate which have occurred during the human period. At a comparatively recent epoch in the record of prehistoric times, we find that the Baltic was far salter than it is now, and produced abundance of oysters; and that Denmark was covered with pine-forests inhabited by capercailzies, such as now only occur farther north in Norway. A little earlier we find that reindeer were common even in the south of France, and still earlier this animal was accompanied by the mammoth and woolly rhinoceros, by the arctic glutton, and by huge bears and lions of extinct species. The presence of such animals implies a change of climate, and both in the caves and gravels we find proofs of a much colder climate than now prevails in Western Europe. Still more remarkable are the changes of the earth's surface which have been effected during man's occupation of it. Many extensive valleys in England and France are believed by the best observers to have been deepened at least a hundred feet; caverns now far out of the reach of any stream must for a long succession of years have had streams flowing through them, at least in times of floods and this often implies that vast masses of solid rock have since been worn away. In Sardinia land has risen at least three hundred feet since men lived there who made pottery and probably used fishing-nets;[3] while in Kent's Cavern remains of man are found buried beneath two separate beds of stalagmite, each having a distinct texture, and each covering a deposit of cave-earth having well-marked differential characters, while each contains a distinct assemblage of extinct animals.

Such, briefly, are the results of the evidence that has been rapidly accumulating for about fifteen years as to the antiquity of man; and it has been confirmed by so many discoveries of a like nature in all parts of the globe, and especially by the comparison of the tools and weapons of prehistoric man with those of modern savages, so that the use of even the rudest flint implements has become quite intelligible, that we can hardly wonder at the vast revolution effected in public opinion. Not only is the belief in man's vast and still unknown antiquity universal among men of science, but it is hardly disputed by any well-informed theologian; and the present generation of science-students must, we should think, be somewhat puzzled to understand what there was in the earliest discoveries that should have aroused such general opposition and been met with such universal incredulity.

But the question of the mere "Antiquity of Man" almost sank into insignificance at a very early period of the inquiry, in comparison with the far more momentous and more exciting problem of the development of man from some lower animal form, which the theories of Mr. Darwin and of Mr. Herbert Spencer soon showed to be inseparably bound up with it. This has been, and to some extent still is, the subject of fierce conflict; but the controversy as to the fact of such development is now almost at an end, since one of the most talented representatives of Catholic theology, and an anatomist of high standing—Prof. Mivart—fully adopts it as regards physical structure, reserving his opposition for those parts of the theory which would deduce man's whole intellectual and moral nature from the same source, and by a similar mode of development.

Never, perhaps, in the whole history of science or philosophy has so great a revolution in thought and opinion been effected as in the twelve years from 1859 to 1871, the respective dates of publication of Mr. Darwin's "Origin of Species" and "Descent of Man." Up to the commencement of this period the belief in the independent creation or origin of the species of animals and plants, and the belief in the very recent appearance of man upon the earth, were, practically, universal. Long before the end of it these two beliefs had utterly disappeared, not only in the scientific world, but almost equally so among the literary and educated classes generally. The belief in the independent origin of man held its ground somewhat longer, but the publication of Mr. Darwin's great work gave even that its death-blow, for hardly any one capable of judging of the evidence now doubts the derivative nature of man's bodily structure as a whole, although many believe that his mind and even some of his physical characteristics may be due to the action of other forces than have acted in the case of the lower animals.

We need hardly be surprised, under these circumstances, if there has been a tendency among men of science to pass from one extreme to the other, from a profession (so few years ago) of total ignorance as to the mode of origin of all living things, to a claim to almost complete knowledge of the whole progress of the universe from the first speck of living protoplasm up to the highest development of the human intellect. Yet this is really what we have seen in the last sixteen years. Formerly difficulties were exaggerated, and it was asserted that we had not sufficient knowledge to venture on any generalizations on the subject. Now difficulties are set aside, and it is held that our theories are so well established and so far-reaching, that they explain and comprehend all Nature. It is not long ago (as I have already reminded you) since facts were contemptuously ignored, because they favored our now popular views; at the present day it seems to me that *facts* which oppose them hardly receive due consideration. And, as opposition is the best incentive to progress, and it is not well even for the best theories to have it all their own way, I propose to direct your attention to a few such facts, and to the conclusions that seem fairly deducible from them.

It is a curious circumstance that, notwithstanding the attention that has been directed to the subject in every part of the world, and the numerous excavations connected with railways and mines which have offered such facilities for geological discovery, no advance whatever has been made for a considerable number of years, in detecting the time or the mode of man's origin. The Palæolithic flint weapons first discovered in the north of France more than thirty years ago are still the oldest undisputed proofs of man's existence; and, amid the countless relics of a former world that have been brought to light, no evidence of any one of the links that must have connected man with the lower animals has yet appeared.

It is, indeed, well known that negative evidence in geology is of very slender value, and this is, no doubt, generally the case. The circumstances here are, however, peculiar; for many converging lines of evidence show that, on the theory of development by the same laws which have determined the development of the lower animals, man must be immensely older than any traces of him yet discovered. As this is a point of great interest, we must devote a few moments to its consideration:

1. The most important difference between man and such of the lower animals as most nearly approach him is undoubtedly in the bulk and development of his brain, as indicated by the form and capacity of the cranium. We should therefore anticipate that these earliest races, who were contemporary with the extinct animals and used rude stone weapons, would show a marked deficiency in this respect. Yet the oldest known crania—those of the Engis and Cro-Magnon caves—show no marks of degradation. The former does not present so low a type as that of most existing savages, but is—to use the words of Prof. Huxley—"a fair average human skull, which might have belonged to a philosopher, or might have contained the thoughtless brains of a savage." The latter are still more remarkable, being unusually large and well formed. Dr. Pruner-Bey states that they surpass the average of modern European skulls in capacity, while their symmetrical forms, without any trace of prognathism, compare favorably not only with, the foremost savage races, but with many civilized nations of modern times.

One or two other crania of much lower type, but of

less antiquity than this, have been discovered; but they in no way invalidate the conclusion which so highly developed a form at so early a period implies, viz., that we have as yet made a hardly perceptible step toward the discovery of any earlier stage in the development of man.

2. This conclusion is supported and enforced by the nature of many of the works of art found even in the oldest cave-dwellings. The flints are of the old chipped type, but they are formed into a large variety of tools and weapons—such as scrapers, awls, hammers, saws, lances, etc.—implying a variety of purposes for which these were used, and a corresponding degree of mental activity and civilization. Numerous articles of bone have also been found, including well-formed needles, implying that skins were sewn together, and perhaps even textile materials woven into cloth. Still more important are the numerous carvings and drawings representing a variety of animals, including horses, reindeer, and even a mammoth, executed with considerable skill on bone, reindeer horns, and mammoth-tusks. These, taken together, indicate a state of civilization much higher than that of the lowest of our modern savages, while it is quite compatible with a considerable degree of mental advancement, and leads us to believe that the crania of Engis and Cro-Magnon are not exceptional, but fairly represent the characters of the race. If we further remember that these people lived in Europe under the unfavorable

conditions of a sub-arctic climate, we shall be inclined to agree with Dr. Daniel Wilson, that it is far easier to produce evidences of deterioration than of progress in instituting a comparison between the contemporaries of the mammoth and later prehistoric races of Europe or savage nations of modern times.[4]

3. Yet another important line of evidence as to the extreme antiquity of the human type has been brought prominently forward by Prof. Mivart.[5] He shows, by careful comparison of all parts of the structure of the body, that man is related, not to any one, but almost equally to many of the existing apes—to the orang, the chimpanzee, the gorilla, and even to the gibbons—in a variety of ways; and these relations and differences are so numerous and so diverse that on the theory of evolution the ancestral form which ultimately developed into man must have diverged from the common stock whence all these various forms and their extinct allies originated. But so far back as the Miocene deposits of Europe, we find the remains of apes allied to these various forms, and especially to the gibbons, so that in all probability the special line of variation which led up to man branched off at a still earlier period. And these early forms, being the initiation of a far higher type, and having to develop by natural selection into so specialized and altogether distinct a creature as man, must have risen at a very early period into the position of a dominant race,

and spread in dense waves of population over all suitable portions of the great continent for this, on Mr. Darwin's hypothesis, is essential to rapid developmental progress through the agency of natural selection.

Under these circumstances we might certainly expect to find some relics of these earlier forms of man along with those of animals which were presumably less abundant. Negative evidence of this kind is not very weighty, but still it has some value. It has been suggested that as apes are mostly tropical, and anthropoid apes are now confined almost exclusively to the vicinity of the equator, we should expect the ancestral forms also to have inhabited these same localities West Africa and the Malay Islands. But this objection is hardly valid, because existing anthropoid apes are wholly dependent on a perennial supply of easily-accessible fruits, which is only found near the equator, while not only had the south of Europe an almost tropical climate in Miocene times, but we must suppose even the earliest ancestors of man to have been terrestrial and omnivorous, since it must have taken ages of slow modification to have produced the perfectly erect form, the short arms, and the wholly non-prehensile foot, which so strongly differentiate man from the apes.

The conclusion which I think we must arrive at is, that if man has been developed from a common ancestor with all existing apes, *and by no other agencies than such as have affected their development*, then he must have existed in something approaching his present form, during the Tertiary period and not merely existed, but predominated in numbers, wherever suitable conditions prevailed. If, then, continued researches in all parts of Europe and Asia fail to bring to light any proofs of his presence, it will be at least a presumption that he came into existence at a much later date, and by a much more rapid process of development. In that case it will be a fair argument that, just as he is in his mental and moral nature, his capacities and aspirations, so infinitely raised above the brutes, so his origin is due to distinct and higher agencies than such as have affected their development.

There is yet another line of inquiry bearing upon this subject to which I wish to call your attention. It is a somewhat curious fact that, while all modern writers admit the great antiquity of man, most of them maintain the very recent development of his intellect, and will hardly contemplate the possibility of men, equal in mental capacity to ourselves, having existed in prehistoric times. This question is generally assumed to be settled by such relics as have been preserved of the manufactures of the older races, showing a lower and lower state of the arts by the successive disappearance in early times of iron, bronze, and pottery; and by the ruder forms of the older flint implements. The weakness of this argument has been well shown by Mr. Albert Mott in his very original but little known presidential address to the

Literary and Philosophical Society of Liverpool in 1873. He maintains that "our most distant glimpses of the past are still of a world peopled as now with men both civilized and savage;" and "that we have often entirely misread the past by supposing that the outward signs of civilization must always be the same, and must be such as are found among ourselves." In support of this view he adduces a variety of striking facts and ingenious arguments, a few of which I will briefly summarize.

On one of the most remote islands of the Pacific—Easter Island—2,000 miles from South America, 2,000 from the Marquesas, and more than 1,000 from the Gambier Islands, are found hundreds of gigantic stone images, now mostly in ruins, often thirty or forty feet high, while some seem to have been much larger, the crowns on their heads cut out of a red stone, being sometimes ten feet in diameter, while even the head and neck of one are said to have been twenty feet high.[6] These once stood erect on extensive stone platforms, yet the island has only an area of about thirty square miles, or considerably less than Jersey. Now, as one of the smallest images eight feet high weighs four tons, the largest must weigh over a hundred tons if not much more; and the existence of such vast works implies a large population, abundance of food, and an established government. Yet how could these coexist in a mere speck of land wholly cut off from the rest of the world? Mr. Mott maintains that this necessarily implies the power of regular communication with larger islands or a continent, the arts of navigation, and a civilization much higher than now exists in any part of the Pacific. Very similar remains in other islands scattered widely over the Pacific add weight to this argument.

The next example is that of the ancient mounds and earthworks of the North American Continent, the bearing of which is even more significant. Over the greater part of the extensive Mississippi Valley four well-marked classes of these earthworks occur. Some are camps, or works of defense, situated on bluffs, promontories, or isolated hills; others are vast inclosures in the plains and lowlands, often of geometric forms, and having attached to them roadways or avenues often miles in length; a third are mounds corresponding to our tumuli, often seventy to ninety feet high, and some of them covering acres of ground; while a fourth group consist of representations of various animals modeled in relief on a gigantic scale, and occurring chiefly in an area somewhat to the northwest of the other classes, in the plains of Wisconsin.

The first class—the camps or fortified inclosures—resemble in general features the ancient camps of our own islands, but far surpass them in extent. Fort Hill, in Ohio, is surrounded by a wall and ditch a mile and a half in length, part of the way cut through solid rock. Artificial reservoirs for water were made within it, while at one extremity, on a more elevated point, a keep is constructed with its separate defenses and water-reservoirs. Another, called Clark's Work, in the Scioto Valley, which seems to have been a fortified town, incloses an area of one hundred and twenty-seven acres, the embankments measuring three miles in length, and containing not less than three million cubic feet of earth. This area incloses numerous sacrificial mounds and symmetrical earthworks in which many interesting relics and works of art have been found.

The second class—the sacred inclosures—may be compared for ex tent and arrangement with Avebury or Carnak—but are in some respects even more remarkable. One of these, at Newark, Ohio, covers an area of several miles with its connected groups of circles, octagons, squares, ellipses, and avenues, on a grand scale, and formed by embankments from twenty to thirty feet in height. Other similar works occur in different parts of Ohio, and by accurate survey it is found not only that the circles are true, though some of them are one-third of a mile in diameter, but that other figures are truly square, each side being over one thousand feet long, and, what is still more important, the dimensions of some of these geometrical figures in different parts of the country, and seventy miles apart, are identical. Now, this proves the use, by the builders of these works, of some standard measures of length, while the accuracy of the squares, circles, and, in a less degree, of the octagonal figures, shows a considerable knowledge of rudimentary geometry, and some means of measuring angles. The difficulty of drawing such figures on a large scale is much greater than any one would imagine who has not tried it, and the accuracy of these is far beyond what is necessary to satisfy the eye. We must therefore impute to these people the wish to make these figures as accurate as possible, and this wish is a greater proof of habitual skill and intellectual advancement than even the ability to draw such figures. If, then, we take into account this ability and this love of geometric truth, and further consider the dense population and civil organization implied by the construction of such extensive systematic works, we must allow that these people had reached the earlier stages of a civilization of which no traces existed among the savage tribes who alone occupied the country when first visited by Europeans.

The animal mounds are of comparatively less importance for our present purpose, as they imply a somewhat lower grade of advancement; but the sepulchral and sacrificial mounds exist in vast numbers, and their partial exploration has yielded a quantity of articles and works of art which throw some further light on the peculiarities of this mysterious people. Most of these mounds contain a large concave hearth or basin of burnt clay, of perfectly symmetrical form, on which are found deposited more or less abundant relics, all bearing traces of the action of fire. We are, therefore, only acquainted with such articles as are practically fire-proof. These consist of bone and copper implements and ornaments, disks, and tubes—pearl, shell, and silver beads, more or less injured by the fire—ornaments cut in mica, ornamental pottery, and numbers of elaborate carvings in stone, mostly forming pipes for smoking. The metallic articles are all formed by hammering, but the execution is very good; plates of mica are found cut into scrolls and circles; the pottery, of which very few remains have been found, is far superior to that of any of the Indian tribes, since Dr. Wilson is of opinion that they must have been formed on a wheel, as they are often of uniform thickness throughout (sometimes not more than one-sixth of an inch), polished and ornamented with scrolls and figures of birds and flowers in delicate relief. But the most instructive objects are the sculptured stone pipes, representing not only various easily-recognizable animals, but also human heads, so well executed that they appear to be portraits. Among the animals, not only are such native forms as the panther, bear, otter, wolf, beaver, raccoon, heron, crow, turtle, frog, rattlesnake, and many others, well represented, but also the manatee, which perhaps then ascended the Mississippi as it now does the Amazon, and the toucan, which could hardly have been obtained nearer than Mexico. The sculptured heads are especially remarkable, because they present to us the features of an intellectual and civilized people. The nose in some is perfectly straight, and neither prominent nor dilated, the mouth is small, and the lips thin, the chin and upper lip are short, contrasting with the ponderous jaw of the modern Indian, while the cheek-bones present no marked prominence. Other examples have the nose somewhat projecting at the apex in a manner quite unlike the features of any American indigenes, and, although there are some which show a much coarser face, it is very difficult to see in any of them that close resemblance to the Indian type which these sculptures have been said to exhibit. The few authentic crania from the mounds present corresponding features, being far more symmetrical and better developed in the frontal region than those of any American tribes, although somewhat resembling them in the occipital outline;[7] while one was described by its discoverer (Mr. W. Marshall Anderson) as "a beautiful skull worthy of a Greek."

The antiquity of this remarkable race may perhaps not be very great, as compared with the prehistoric man of Europe, although the opinions of some writers on the subject seem affected by that "parsimony of time" on which the late Sir Charles Lyell so often dilated. The mounds are all overgrown with dense forest, and one of the large trees was estimated to be 800 years old, while other observers consider the forest-growth to indicate an age of at least 1,000 years. But it is well known that it requires several generations of trees to pass away before the growth on a deserted clearing comes to correspond with that of the surrounding virgin forest, while this forest, once established, may go on growing for an unknown number of thousands of years. The 800 or 1,000 years estimate from the growth of existing vegetation is a minimum which has no bearing whatever on the actual age of these mounds, and we might almost as well attempt to determine, the time of the glacial epoch from the age of the pines or oaks which now grow on the moraines.

The important thing for us, however, is, that when North America was first settled by Europeans, the Indian tribes inhabiting it had no knowledge or tradition of any preceding race of higher civilization than themselves. Yet we find that such a race existed; that they must have been populous, and have lived under some established government; while there are signs that they practised agriculture largely, as indeed they must have done to have supported a population capable of executing such gigantic works in such vast profusion—for it is stated that the mounds and earthworks of various kinds in the State of Ohio alone amount to between eleven and twelve thousand. In their habits, customs, religion, and arts, they differed strikingly from all the Indian tribes; while their love of art and geometric forms and their capacity for executing the latter upon so gigantic a scale render it probable that they were a really civilized people, although the form their civilization took may have been very different from that of later people subject to very different influences, and the inheritors of a longer series of ancestral civilizations. We have here, at all events, a striking example of the transition, over an extensive country, from comparative civilization to comparative barbarism, the former having left no tradition, and hardly any trace of influence on the latter.

As Mr. Mott well remarks: "Nothing can be more striking than the fact that Easter Island and North America both give the same testimony as to the origin of the savage life found in them, although in all circumstances and surroundings the two cases are so different. If no stone monuments had been constructed in Easter Island, or mounds, containing a few relics saved from fire, in the United States, we might never have suspected the existence of these ancient peoples." He argues, therefore, that it is very easy for the records of an ancient nation's life entirely to perish, or to be hidden from observation. Even the arts of Nineveh and Babylon were unknown only a generation ago, and we have only just discovered the facts about the mound-builders of North America.

But other parts of the American Continent exhibit parallel phenomena. Recent investigations show that in Mexico, Central America, and Peru, the existing race of Indians has been preceded by a distinct and more civilized race. This is proved by the sculptures of the ruined cities of Central America, by the more ancient *terra-Cottas* and paintings of Mexico, and by the oldest portrait-pottery of Peru. All alike show markedly non-Indian features, while they often closely resemble modern European types. Ancient crania, too, have been found in all these countries, presenting very different characters from those of any of the modern indigenous races of America.[8]

There is one other striking example of a higher being succeeded by a lower degree of knowledge, which is in danger of being forgotten because it has been made the foundation of theories which seem wild and fantastic, and are probably in great part erroneous. I allude to the Great Pyramid of Egypt, whose form, dimensions, structure, and uses, have recently been the subject of elaborate works by Prof. Piazzi Smyth. Now, the admitted facts about this pyramid are so interesting and so apposite to the subject we are considering, that I beg to recall them to your attention. Most of you are aware that this pyramid has been carefully explored and measured by successive Egyptologists, and that the dimensions have lately become capable of more accurate determination, owing to the discovery of some of the original casing-stones and the clearing away of the earth from the corners of the foundation, showing the sockets in which the corner-stones fitted. Prof. Smyth devoted many months of work with the best instruments in order to fix the dimensions and angles of all accessible parts of the structure; and he has carefully determined these by a comparison of his own and all previous measures, the best of which agree pretty closely with each other. The results arrived at are:

1. That the pyramid is truly square, the sides being equal and the angles right angles.

2. That the four sockets on which the first four stones of the corners rested are truly on the same level.

3. That the directions of the sides are accurately to the four cardinal points.

4. That the vertical height of the pyramid bears the same proportion to its circumference at the base as the radius of a circle does to its circumference.

Now all these measures, angles, and levels, are accurate, not as an ordinary surveyor or builder could make them, but to such a degree as requires the very best modern instruments and all the refinements of geodetical science to discover any error at all. In addition to this we have the wonderful perfection of the workmanship in the interior of the pyramid, the passages and chambers being lined with huge blocks of stones fitted with the utmost accuracy, while every part of the building exhibits the highest structural science.

In all these respects this largest pyramid surpasses every other in Egypt. Yet it is universally admitted to be the oldest, and also the oldest historical building in the world.

Now these admitted facts about the Great Pyramid are surely remarkable, and worthy of the deepest consideration. They are facts which, in the pregnant words of the late Sir

John Herschel, "according to received theories ought not to happen," and which, he tells us, should therefore be kept ever present to our minds, since "they belong to the class of facts which serve as the clew to new discoveries." According to modern theories, the higher civilization is ever a growth and an outcome from a preceding lower state; and it is inferred that this progress is visible to us throughout all history and in all the material records of human intellect. But here we have a building which marks the very dawn of history—which is the oldest authentic monument of man's genius and skill, and which, instead of being far inferior, is very much superior to all which followed it. Great men are the products of their age and country, and the designer and constructors of this wonderful monument could never have arisen among an unintellectual and half-barbarous people. So perfect a work implies many preceding less perfect works which have disappeared. It marks the culminating point of an ancient civilization, of the early stages of which we have no record whatever.

The three cases to which I have now adverted (and there are many others) seem to require for their satisfactory interpretation a somewhat different view of human progress from that which is now generally accepted. Taken in connection with the great intellectual power of the ancient Greeks—which Mr. Galton believes to have been far above that of the average of any modern nation—and the elevation, at once intellectual and moral, displayed in the writings of Confucius, Zoroaster, and the Vedas, they point to the conclusion that, while in material progress there has been a tolerably steady advance, man's intellectual and moral development reached almost its highest level in a very remote past. The lower, the more animal, but often the more energetic types, have, however, always been far the more numerous; hence such established societies as have here and there arisen under the guidance of higher minds have always been liable to be swept, away by the incursions of barbarians. Thus, in almost every part of the globe there may have been a long succession of partial civilization, each in turn succeeded by a period of barbarism; and this view seems supported by the occurrence of degraded types of skull along with such "as might have belonged to a philosopher"—at a time when the mammoth and the reindeer inhabited Southern France.

Nor need we fear that there is not time enough for the rise and decay of so many successive civilizations as this view would imply; for the opinion is now gaining ground among geologists that paleolithic man was really preglacial, and that the great gap—marked alike by a change of physical conditions, and of animal life—which in Europe always separates him from his neolithic successor, was caused by the coming on and passing away of the great Ice age.

If the views now advanced are correct, many, perhaps most, of our existing savages are the successors of higher races; and their arts, often showing a wonderful similarity in distant continents, may have been derived from a common source among more civilized peoples.

I must now conclude this very imperfect sketch of a few of the offshoots from the great tree of biological study. It will, perhaps, be thought by some that my remarks have tended to the depreciation of our science, by hinting at imperfections in our knowledge and errors in our theories, where more enthusiastic students see nothing but established truths. But I trust that I may have conveyed to many of my hearers a different impression. I have endeavored to show that even in what are usually considered the more trivial and superficial characters presented by natural objects, a whole field of new inquiry is opened up to us by the study of distribution and local conditions. And as regards man, I have endeavored to fix your attention on a class of facts which indicate that the course of his development has been far less direct and simple than has hitherto been supposed; and that, instead of resembling a single tide with its advancing and receding ripples, it must rather be compared to the progress from neap to spring tides, both the rise and the depression being comparatively greater as the waters of true civilization slowly advance toward the highest level they can reach.

And if we are thus led to believe that our present knowledge of Nature is somewhat less complete than we have been accustomed to consider it, this is only what we might expect; for, however great may have been the intellectual triumphs of the nineteenth century, we can hardly think so highly of its achievements as to imagine that, in somewhat less than twenty years, we have passed from complete ignorance to almost perfect knowledge on two such vast and complex subjects as the origin of species and the antiquity of man.

⊠ From the opening address of Mr. Wallace, as President of the Biological Section of the British Association for the Advancement of Science, given at its recent meeting in Glasgow.

⊠ In 1854 (?) a communication from the Torquay Natural History Society, confirming previous accounts by Mr. Godwin-Austen, Mr. Vivian, and the Rev. Mr. McEnery, that worked flints occurred in Kent's Hole, with remains of extinct species, was rejected as too improbable for publication.

⊠ Lyell's "Antiquity of Man," fourth edition, p. 115.

⊠ "Prehistoric Man," third edition, vol. i., p: 117.

⊠ "Man and Apes," pp. 171-193.

⊠ Journal of the Royal Geographical Society, 1870, pp.

177, 178.

☒ Wilson's "Prehistoric Man," third edition, vol. ii., pp. 123-130.

☒ Wilson's "Prehistoric Man," third edition, vol. ii., pp. 125, 144.